Firearms Commerce
in the
United
States

2001/2002

Department of the Treasury
Bureau of Alcohol, Tobacco & Firearms

*"Working for a Sound and Safer America
through Innovation and Partnership"*

Firearms Commerce in the United States

Table of Contents

Introduction   1

Part I - Manufacturers' Firearms Entering Into Commerce.   3

    Manufacturers' Sales, Exports, and Imports   3

    Regulation of Firearms Importation and Exportation   3

    Statutory Requirements for Importation.   4

    Technical Assistance to Manufacturers and Importers   4

Part II – Regulatory Initiatives   5

    License Renewal Procedures.   5

    Access 2000.   6

    FFL eZ Check System   6

    Industry Education and Partnerships   7

Afterword.   8

Exhibits

1. Firearms Manufactured (1986-1999)   E-1

2. Firearms Exports (1986 – 2000)   E-2

3. Firearms Imports (1986-2000)   E-3

4. Importation Applications (FY 1986 – FY 2001)   E-4

5. Firearms Imported into the United States by Country (2000)   E-5

6. Firearms and Ammunition Excise Tax Collections (1983-2000)   E-6

7. National Firearms Act Weapons Registered, Tax Revenues, and Related Activities (1990-2000)   E-7

8. National Firearms Act Applications (1990 – 2000)   E-8

9. National Firearms Act Registered Weapons by State (2000)   E-9

10. National Firearms Act Special Occupational Taxpayers (1980 – 2001)   E-10

11. National Firearms Act Special Occupational Taxpayers (as of January 2001) by State   E-11

12. Federal Firearms Licensees Total (FY 1975 – FY 2001)   E-12

13. Federal Firearms Licensees by State, Number, and Rate per 100,000 Population (2001)   E-13

14. License Applications and Application Inspections (FY 1990 – FY 2001)   E-14

15. Actions on Federal Firearms License Applications (FY 1975 – FY 2001)   E-15

16. Federal Firearms Licensees and Compliance Inspections (FY 1969 – FY 2001)   E-16

INTRODUCTION

This report, the second in what will be an annual series, provides an overview of firearms commerce as it affects the United States. As noted in the first edition, <u>Commerce in Firearms in the United States: February 2000</u>, the Bureau of Alcohol, Tobacco and Firearms (ATF) is responsible for enforcing the Gun Control Act of 1968 (GCA). The GCA regulates the manufacture, importation, distribution, and sale of firearms, and it contains criminal provisions related to the illegal possession, use, or sale of firearms. ATF also administers the National Firearms Act (NFA), which requires the registration of certain weapons, such as machineguns and destructive devices, and imposes taxes on the making and transfer of those weapons.

This edition contains information about domestic firearms manufacturing, as well as about the importation and exportation of firearms. It also provides an update of certain Bureau regulatory initiatives, such as efforts to improve regulatory compliance and the firearms tracing process, and the fostering of partnerships with the firearms industry.

Part I contains information about domestic firearms manufacturing and the importation and exportation of firearms, which is intended to provide a better understanding of the firearms market.

Part II highlights initiatives that the Bureau has developed to improve the integrity of the firearms license renewal process, to facilitate crime gun tracing,[1] and to prevent illegal firearms sales and trafficking. It also discusses ATF's efforts to foster partnerships with the firearms industry.

The appendix to this report is a series of statistical tables, which contain the most up-to-date information available about the firearms industry and ATF's regulatory activities. Through its regulatory activities, ATF works to ensure compliance with the Federal firearms laws and to prevent firearms from either knowingly or inadvertently being diverted to individuals prohibited from possessing them.

---

[1] Crime gun tracing is the systematic tracking of the movement of a firearm recovered by law enforcement officials from its first sale by the manufacturer or importer through the distribution chain to the first retail purchaser. The tracing process enables law enforcement officials to develop investigative leads, identifying suspects or individuals from whom crime guns are obtained.

PART I

Manufacturers' Firearms Entering Into Commerce

In enforcing the Gun Control Act, the National Firearms Act, and the firearms excise tax provisions of the Internal Revenue Code, ATF collects information on the manufacture, importation, and exportation of firearms. This section presents information on current firearms manufacturers and their reported sales, along with a discussion of the procedures to import and export firearms.

Manufacturers' Sales, Exports, and Imports

There are currently more than 1,700 licensed firearms manufacturers and nearly 750 licensed importers in the United States. A Federal firearms license is required to engage in the business of manufacturing or importing firearms. These businesses are required by law to maintain records of the production, exportation, and importation of firearms.

Manufacturers' reports to ATF show the number of manufactured firearms "disposed of in commerce" each calendar year, as well as the number produced for export. The term "disposed of in commerce" refers to manufacturers' final sales, which equal the number of firearms produced minus the increase in manufacturers' inventories of firearms. In 1999, the most recent year for which information is available, manufacturers disposed of more than 1.3 million handguns and nearly 2.6 million rifles and shotguns in commerce. Detailed production information from these reports can be found in the appendix to this report.

Importation and exportation statistics may also be found in the appendix to this report. Trends show that imports over the past 10 years have averaged 1 million per year, while exports have averaged less than 400,000 per year during the same time period.

Regulation of Firearms Importation and Exportation

The process of importing and exporting firearms to and from the United States is overseen by several Federal agencies. ATF administers the import provisions of the Gun Control Act, the National Firearms Act, and the Arms Export Control Act. This includes the approval or denial of applications to import firearms and ammunition for persons, businesses and governmental entities who wish to import such materials into the United States.

ATF also provides technical advice and assistance to the general public regarding import requirements applicable to any firearms or ammunition that they are bringing into the United States from another country (see "Technical Assistance to Manufacturers and Importers" below).

The Office of Defense Trade Controls within the Department of State regulates the exportation of firearms other than sporting shotguns. A person wishing to export such a firearm must obtain an exportation license from the Department of State prior to the firearm's shipment. Within the Department of Commerce, the Bureau of Export

Administration regulates the exportation of sporting shotguns with barrels between 18 and 28 inches in length. A general license is required to export sporting shotguns.

Statutory Requirements for Importation

The Gun Control Act of 1968 (GCA) generally prohibits the importation of firearms into the United States. However, the GCA creates four categories of firearms that the Secretary of the Treasury must authorize for importation: These include firearms that are (1) being used for scientific or research purposes, or particular competition or training purposes, (2) unserviceable firearms (other than machineguns) that are designated as curios

or museum pieces, (3) firearms that were previously taken out of the United States by the person who is bringing them back, and (4) firearms--other than National Firearms Act and surplus military weapons--that are of a type "generally recognized as particularly suitable for or readily adaptable to sporting purposes" (the "sporting purposes" test). Firearms in this final category comprise the majority of those that are imported into the United States.

Technical Assistance to Manufacturers and Importers

ATF maintains a staff of firearms technology specialists who provide technical advice and services for manufacturers and importers of firearms.

These specialists conduct examinations of firearms and related products, including industry prototypes, and make determinations on the importability of certain firearms (e.g., whether the firearm meets the sporting purposes test).

The staff also classifies weapons in order to support law enforcement investigations and programs. Weapons classifications include machineguns, silencers, semiautomatic assault weapons, and a wide assortment of pistols, revolvers, rifles, shotguns, and firearms-related accessories. Specialists also respond to numerous inquiries from manufacturers and importers, other Federal firearms licensees, and the general public.

PART II

Regulatory Initiatives

Part II discusses recently completed or ongoing regulatory initiatives that have improved the Bureau's ability to ensure compliance by the firearms industry and to prevent illegal firearms sales and trafficking.

License Renewal Procedures

ATF is responsible for ensuring that only individuals who are engaged in the firearms business are permitted to hold Federal firearms licenses. ATF conducted an inspection program in 1993 entitled Operation Snapshot, which found that approximately 46 percent of licensed dealers conducted no business but rather used their licenses to buy firearms across State lines at wholesale prices. Further, Federal licensing procedures at the time did not take into account whether the licensees were violating State or local zoning laws.

In 1993 and 1994, Congress placed new restrictions on potential licensees, including increasing the licensing fee for dealers and pawnbrokers from $10 per year to $200 for the first 3 years and $90 for each 3-year renewal; requiring applicants to notify the chief of their local law enforcement agency of their intent to apply for a license; and requiring that individuals submit photos and fingerprints with their applications, as well as a certification that their business is in compliance with all State and local laws. ATF now evaluates this information to determine applicants' eligibility for licensing.

The licensee population decreased from a high of more than 286,000 in April 1993 to a low of 102,020 in March 2000, largely as a result of these changes.

Figure 1, which follows, gives an overview of the licensee population between calendar years 1990 and 2001. Exhibit 12 in the appendix provides an overview of specific categories of firearms licensees.

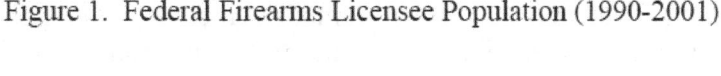
Figure 1. Federal Firearms Licensee Population (1990-2001)

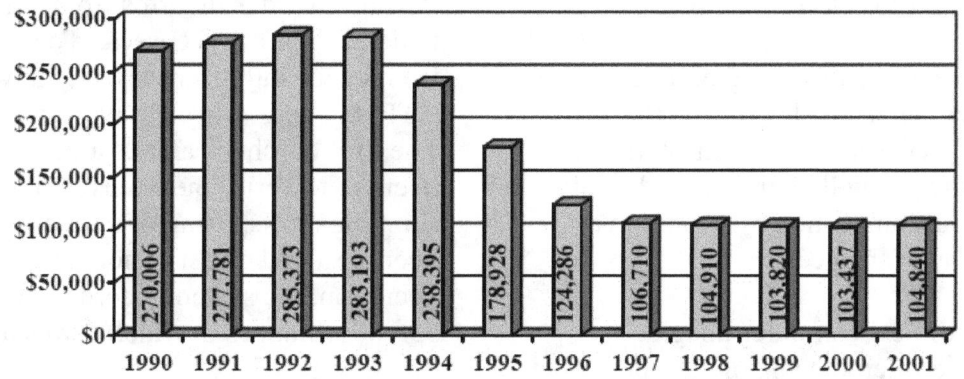

In 1998, the Bureau conducted a second Operation Snapshot program, which showed that 31 percent of licensed dealers had not documented a firearms sale within the previous year. Although this figure had dropped from 1993, additional measures were clearly required to ensure the integrity of the firearms industry.

In August 2000, ATF revised its Federal firearms license renewal application to require firearms dealers to document the number of firearms they have acquired and disposed of over the prior 3 years. With this change, the Bureau is able to more effectively determine whether renewal applicants are engaging in the firearms business and whether their licenses should be renewed.

This requirement does not apply to gunsmiths who do not sell firearms or to collectors of curio and relic firearms.

Access 2000

ATF continues to improve the efficiency of the firearms tracing process while reducing costs and burdens on the firearms industry with its Access 2000 initiative. Access 2000 provides an electronic link for the Bureau's National Tracing Center to access the firearms records of participating manufacturers, importers, and wholesalers when it requires information to complete trace requests.

The system does not provide the National Tracing Center with information about individual purchasers, nor does it allow the Center to examine all the distribution records of the participant at once. It limits ATF personnel to conducting single queries about individual crime guns. However, Access 2000 does allow the National Tracing Center to make queries 24 hours a day, 7 days a week. It reduces the burden on licensees, who no longer have to provide personnel to examine their own records to discover how crime guns have been distributed. Currently, 12 licensees are participating in the project, and ATF is working to expand licensee participation.

FFL eZ Check System

The "FFL eZ Check," which became operational in October 2000, was created to help the firearms industry prevent the fraudulent use of firearms licenses.

Prior to a licensee's disposing of a firearm to another licensee, he or she must verify the identity and licensed status of the person to whom the firearm will be transferred. This is generally accomplished by obtaining a certified copy of the license.

The advent of new computer imaging, scanning, and Internet technologies has made it increasingly easier for an unlicensed individual to create an authentic looking copy of a license and to use that copy to attempt to order firearms from legitimate Federal firearms licensees. To prevent this from occurring, licensees may now access ATF's FFL eZ Check at www.atf.treas.gov. The FFL eZ check allows a Federal firearms licensee to verify the license prior to shipping or disposing of a firearm to a licensee. In the first 6 months of its operation, the system recorded more than 55,000 instances in which it was used.

The system also provides a toll-free number (1-877-560-2435) for licensees to query to determine the validity of a particular Federal firearms license. The toll-free number is operational 7 days a week during customary working hours.

Industry Education and Partnerships

ATF is reaching out to strengthen and develop new working relationships with the firearms industry and consumers through education and partnerships. To this end, the Bureau regularly holds informational seminars for licensed dealers to keep them informed about legislative or regulatory changes that will affect their businesses. ATF also provides a variety of instructional and informational materials to the industry, including regularly updated reference guides to the Federal firearms laws and regulations, licensee newsletters, and a manual for dealers to evaluate their vulnerability to thefts. Additional information is also provided through ATF's website at www.atf.treas.gov.

To address industry concerns, ATF has established regularly scheduled meetings with representatives from the National Coalition of Firearms Retailers, the National Shooting Sports Foundation (NSSF) and the Sporting Arms and Ammunition Manufacturer's Association to deal with firearms issues that affect industry and the general public. ATF and NSSF have partnered to create a two-pronged educational campaign, entitled "Don't Lie for the Other Guy," in an effort to prevent prohibited individuals from obtaining firearms. This campaign is intended to discourage people from illegally purchasing firearms on behalf of others, often for individuals who are unable to legally possess them. Moreover, it heightens awareness of these illegal sales among licensed dealers.

The first element of the campaign involves mailing kits to firearms retailers that contain posters, pamphlets, and other printed material provided by NSSF. The second involves ATF and NSSF jointly conducting a series of educational seminars in several locations for Federal firearms licensees.

ATF also meets frequently with representatives of the Firearms Importers Round Table trade group to discuss issues of mutual interest or concern relating to the importation of firearms and other munitions.

Afterword

ATF strives to provide the public with the most up-to-date information on the firearms industry, as well as to promote compliance by, and partnerships with, the industry. The Bureau will continue to publish yearly statistical data relating to the industry and information concerning new regulatory initiatives. Any suggestions to improve the usefulness of the data provided in this report may be submitted to ATF, Attn: Firearms Programs Division, 650 Massachusetts Avenue NW., Washington, DC 20226.

Exhibit 1. Firearms Manufactured (1986-1999) [a]

Calendar Year	Pistols	Revolvers	Rifles	Shotguns	Machineguns	Total Firearms
1986	692,977	734,650	970,541	641,482	41,482	3,081,132
1987	963,562	695,270	1,006,100	857,949	3,963	3,526,844
1988	991,011	754,711	1,144,707	928,070	2,239	3,820,738
1989	1,402,660	628,765	1,407,317	935,541	2,387	4,376,670
1990	1,376,399	462,496	1,156,213	848,948	3,809	3,847,865
1991	1,381,325	456,941	883,482	828,426	2,213	3,552,387
1992	1,549,659	460,373	1,001,708	1,018,204	900	4,030,844
1993	2,272,001	552,808	1,160,124	1,144,940	4,240	5,134,113
1994	1,995,511	586,450	1,324,240	1,254,926	10,248	5,171,375
1995	1,195,266	527,664	1,331,780	1,173,645	9,185	4,237,540
1996	985,533	498,944	1,424,319	925,732	22,020	3,856,548
1997	1,036,077	370,428	1,251,341	915,978	67,844	3,641,668
1998	916,070	324,390	1,535,690	868,639	32,866	3,677,655
1999	995,446	335,784	1,569,685	1,106,995	22,490	4,030,400

Source: Bureau of Alcohol, Tobacco and Firearms' Annual Firearms Manufacturing and Exportation Report.

[a] The manufacturers' reports exclude production for the U.S. military but include firearms purchased by domestic law enforcement agencies. They also include firearms manufactured for export.

Exhibit 2. Firearms Exports (1986 - 2000)

Manufacturers' Exports (1986 - 1999)

Calendar Year	Pistols	Revolvers	Rifles	Shotguns	Machineguns	Total Firearms
1986	16,657	103,890	37,224	58,943	24,781	241,495
1987	24,941	133,859	42,144	41,014	24,448	266,406
1988	32,570	99,289	53,896	68,699	12,338	266,792
1989	41,976	76,494	73,247	67,559	11,599	270,875
1990	73,398	104,620	71,659	104,250	19,337	373,264
1991	79,462	110,058	91,111	117,801	36,785	435,217
1992	77,309	111,821	89,965	119,127	10,219	408,441
1993	59,080	89,641	94,170	171,475	7,012	421,378
1994	93,956	78,935	81,835	146,524	16,717	417,967
1995	97,969	131,634	89,053	100,894	19,259	438,809
1996	64,126	90,058	74,555	97,173	33,875	359,787
1997	44,182	63,656	76,626	86,263	20,857	291,584
1998	28,805	15,788	65,807	89,699	12,529	212,628
1999	34,663	48,616	65,669	67,342	22,255	238,545

Source: Bureau of Alcohol, Tobacco and Firearms' Annual Firearms Manufacturing and Exportation Report.

Total U.S. Exports (1997 - 2000)

Calendar Year	Revolvers & Pistols	Centerfire Rifles (sporting)	Rimfire Rifles (sporting)	Shotguns	Combination Shotgun/Rifle
1997	148,023	73,365	33,442	104,562	1,773
1998	124,295	49,890	35,860	127,641	8,521
1999	116,467	48,713	20,676	77,583	1,861
2000	82,185	49,081	18,132	95,244	5,449

Source - U.S. Bureau of the Census: Foreign Trade Division.

Total U.S. Exports of Military Firearms (1997 - 2000)[a]

Calendar Year	Military Rifles	Military Shotguns	Military Machineguns
1997	149,343	3,929	3,079
1998	105,159	4,561	6,434
1999	88,354	3,239	10,302
2000	99,700	4,733	1,892

Source - U.S. Bureau of the Census: Foreign Trade Division.

[a] Military firearms exclude those exported to the U.S. Armed Forces.

E-2

Exhibit 3. Firearms Imports (1986 - 2000)

| Year | Imported Firearms | | | |
	Shotguns	Rifles	Handguns	Total
1986	201,000	269,000	231,000	701,000
1987	307,620	413,780	342,113	1,063,513
1988	372,008	282,640	621,620	1,276,268
1989	274,497	293,152	440,132	1,007,781
1990	191,787	203,505	448,517	843,809
1991	116,141	311,285	293,231	720,657
1992 [a]	441,933	1,423,189	981,588	2,846,710
1993	246,114	1,592,522	1,204,685	3,043,321
1994	117,866	847,868	915,168	1,880,902
1995	136,126	261,185	706,093	1,103,404
1996	128,456	262,568	490,554	881,578
1997	106,296	358,937	474,182	939,415
1998	219,387	248,742	531,681	999,810
1999	385,556	198,191	308,052	891,799
2000	331,985	298,894	465,903	1,096,782

Source: Bureau of Alcohol, Tobacco and Firearms and U.S. Customs Service.

[a] Statistics prior to 1992 are for fiscal years; statistics after 1992 are for calendar years; and 1992 is a transition year with five quarters.

Exhibit 4. Importation Applications (FY 1986 - FY 2001)

| Fiscal Year | Total | Applications for Importation (Forms 6) Processed [a] | | |
		Licensed Importer	Military	Other
1986	19,793	7,728	9,434	2,631
1987	18,022	7,833	8,059	2,130
1988	17,513	7,711	7,680	2,122
1989	18,437	7,950	8,293	2,194
1990	19,248	8,292	8,696	2,260
1991	21,483	8,098	10,973	2,412
1992	19,805	7,960	9,222	2,623
1993	16,458	7,591	6,282	2,585
1994	14,298	6,704	4,570	3,024
1995	10,649	5,267	2,834	2,548
1996	11,527	6,340	2,792	2,395
1997	11,752	8,288	2,069	1,395
1998	13,019	8,767	2,715	1,536
1999	12,776	9,505	2,235	1,036
2000	12,135	7,834	2,885	1,416
2001	15,192	9,639	3,984	1,569

Source: Bureau of Alcohol, Tobacco and Firearms, Firearms and Explosives Imports Data Base.

[a] ATF Form 6 is both the application and permit to import firearms, ammunition, and implements of war. An importer can include numerous articles on one Form 6. The figures above reflect the numbers of applications processed, not the numbers of articles approved.

Exhibit 5. Firearms Imported into the United States by Country (2000)

Country of Export	Shotguns	Rifles	Handguns	Total Firearms
Austria	7	9,328	245,869	255,204
Brazil	55,605	31,486	160,548	247,639
Italy	171,536	10,911	41,325	223,772
Germany	5,976	7,655	114,876	128,507
Canada	3	87,250	15,151	102,404
Japan	7,070	57,163	0	64,233
Russia	39,683	9,603	1,150	50,436
Czech Republic	115	19,425	19,006	38,546
Argentina	0	0	32,237	32,237
Turkey	30,308	0	0	30,308
Belgium	519	22,067	2,004	24,590
Romania	0	21,060	2,000	23,060
Israel	23	125	22,373	22,521
Finland	0	18,954	0	18,954
Spain	6,382	2	11,935	18,319
Hungary	13	3,100	12,200	15,313
Philippines	282	1,356	11,310	12,948
United Kingdom	4539	5169	3	9,711
China[a]	9,320	0	0	9,320
Switzerland	2	2,178	6,118	8,298
Bulgaria	0	300	6,429	6,729
Croatia	0	0	5,500	5,500
France	98	396	1,738	2,232
Sweden	511	694	0	1,205
All Others[b]	691	45	890	1,626
Totals:	331,985	298,894	465,903	1,096,782

Source: U.S. International Trade Commission.

[a] On May 26, 1994, the United States instituted a firearms imports embargo against China. Shotguns, however, are exempt from the embargo.

[b] Imports of fewer than 1,000 per country.

Exhibit 6. Firearms and Ammunition Excise Tax Collections (1983-2000)[a]

Fiscal Year	Total	Pistols and revolvers	Other Firearms	Shells and cartridges
		Dollars in Thousands		
1983	$90,637	$24,080	$34,711	$31,846
1984	$87,665	$22,011	$37,276	$28,378
1985	$102,403	$25,107	$48,906	$28,390
1986	$98,362	$23,433	$39,037	$35,892
1987	$102,521	$25,361	$42,182	$34,978
1988	$114,064	$29,074	$48,867	$36,123
1989	$134,277	$38,230	$48,870	$47,177
1990	$137,409	$42,015	$61,402	$33,992
1991	$144,745	$42,226	$50,237	$52,282
1992	$140,608	$41,760	$45,697	$53,151
1993	$171,434	$54,019	$60,482	$56,933
1994	$213,966	$68,533	$75,637	$69,796
1995	$184,302	$53,779	$72,947	$57,576
1996	$157,816	$38,649	$72,422	$46,745
1997 [b]	$149,090	...	...	...
1998	$164,789	...	...	...
1999	$187,977	...	...	...
2000	$197,840	...	...	...

Source: Bureau of Alcohol, Tobacco and Firearms. The tax rates on the displayed categories are as follows: pistols and revolvers, 10 percent of sale price; firearms other than pistols and revolvers, 11 percent of sale price; shells and cartridges, 11 percent of sale price.

[a] Until FY 1990, the Internal Revenue Service collected excise taxes. ATF assumed the collection in FY 1991.

[b] ATF no longer maintains these statistics by individual category.

E-6

Exhibit 7. National Firearms Act Weapons Registered, Tax Revenues, and Related Activities (1990 - 2000) [a]

Year [b]	Number of weapons registered [c]	Tax revenues ($ in thousands) [d]		Enforcement Support [e]	
		Occupational tax	Transfer and making tax	Certifications	Records checks
1990	439,339	$1,442	$1,308	666	7,981
1991	477,020	$1,556	$1,210	764	7,857
1992	538,875	$1,499	$1,237	1,257	8,582
1993	613,079	$1,493	$1,264	1,024	7,230
1994	678,077	$1,444	$1,596	586	6,283
1995	756,260	$1,007	$1,311	882	5,677
1996	823,459	$1,143	$1,402	529	5,215
1997	905,647	$1,284	$1,630	488	4,395
1998	1,016,863	$1,299	$1,969	353	3,824
1999	1,148,984	$1,330	$2,422	345	3,994
2000	1,271,568	$1,399	$2,301	422	4,690

Source: Bureau of Alcohol, Tobacco and Firearms, National Firearms Registration and Transfer Record (NFRTR).

[a] National Firearms Act weapons, which are defined in the Internal Revenue Code, 26 U.S.C. Chapter 53, include items such as machineguns, short-barreled rifles and shotguns, and destructive devices.

[b] Data from 1990 - 1996 are on a fiscal year basis; data for 1997 - 2000 represent calendar years.

[c] Data represents the cumulative number of weapons registered.

[d] Importers, manufacturers, or dealers in NFA firearms are subject to a yearly occupational tax. There is an additional tax for each firearm transferred or made, generally $200 per weapon. Occupational tax revenues for FY 1990 - 1996 include collections made during the fiscal year for prior tax years.

[e] ATF searches the NFRTR in support of criminal investigations and regulatory inspections in order to determine whether individuals are legally in possession of NFA weapons and whether transfers are made appropriately.

Exhibit 8. National Firearms Act Applications (1990 - 2000) [a]

	Transfers		Personal/Government
	Application for taxpaid transfer	Application for tax-exempt transfer [b]	application to make NFA firearms [c]
Year	(ATF Form 4)	(ATF Form 5)	(ATF Form 1)
1990	7,024	54,959	399
1991	5,395	44,146	524
1992	6,541	45,390	351
1993	7,388	60,193	310
1994	7,600	67,580	1,076
1995	8,263	60,055	1,226
1996	6,418	72,395	1,174
1997	7,873	70,690	855
1998	10,181	93,135	1,093
1999	11,768	95,554	1,071
2000	11,246	96,234	1,334

	Manufactured and imported	Exported
Year	(ATF Form 2)	(ATF Form 9)
1990	66,084	21,725
1991	80,619	40,387
1992	107,313	22,120
1993	70,342	24,041
1994	97,665	34,242
1995	95,061	31,258
1996	103,511	40,439
1997	110,423	36,284
1998	141,101	40,221
1999	137,373	28,128
2000	141,763	28,672

	Tax exempt licensees	Number of
Year	(ATF Form 3)	transactions
1990	23,149	194,215
1991	19,507	201,391
1992	26,352	169,762
1993	22,071	221,627
1994	27,950	238,945
1995	18,593	216,026
1996	16,931	242,054
1997	18,371	246,781
1998	27,921	315,641
1999	28,288	306,515
2000	23,335	309,006

Source: Bureau of Alcohol, Tobacco and Firearms, NFA Special Taxpayers and Revenue Collected Database.

[a] Data from 1990 - 1996 are on a fiscal year basis; data for 1997 - 2000 represent calendar years.

[b] Firearms may be transferred to the U.S. Government or its possessions, to State governments, or to official police organizations without the payment of a transfer tax. Further, transfers of NFA firearms between licensees registered as importers, manufacturers, or dealers who have paid the special occupational tax are likewise exempt from taxation.

[c] Firearms manufactured by, or on behalf of, the U.S. Government or any department, independent establishment, or agency thereof are exempt from the making tax.

E-8

Exhibit 9. National Firearms Act Registered Weapons by State (2000)

State	Total	Machinegun	Silencer	Short-barreled Rifle	Short-barreled Shotgun	Destructive Device [a]	Any Other Weapon [b]	Other [c]
Alabama	27,478	10,084	2,982	392	1,271	11,774	965	10
Alaska	4,370	1,337	501	50	467	1,733	281	1
Arizona	14,524	2,888	1,408	156	635	8,973	443	21
Arkansas	59,152	10,705	4,638	852	1,002	41,185	748	22
California	122,083	14,916	1,869	1,136	5,400	95,222	3,464	76
Colorado	22,566	3,717	1,157	407	893	15,604	770	18
Connecticut	26,463	17,289	2,134	379	1,446	4,570	624	21
Delaware	1,080	146	22	31	212	639	30	0
District of Columbia	10,491	2,153	77	40	377	7,784	60	0
Florida	75,424	15,285	7,515	523	2,212	47,591	2,280	18
Georgia	43,849	14,167	9,547	435	6,156	12,110	1,400	34
Hawaii	2,302	246	20	41	42	1,916	35	2
Idaho	9,077	2,203	1,257	187	223	4,722	478	7
Illinois	32,866	9,310	344	393	1,336	20,534	920	29
Indiana	35,886	10,895	3,191	257	4,057	16,510	952	24
Iowa	8,824	1,277	91	175	622	5,784	851	24
Kansas	11,895	1,579	98	191	618	8,749	641	19
Kentucky	19,033	4,568	1,609	249	1,064	10,874	653	16
Louisiana	29,724	3,602	879	187	835	23,725	481	15
Maine	5,493	2,541	497	466	287	1,069	633	
Maryland	33,985	8,346	2,316	266	1,741	20,586	717	13
Massachusetts	11,386	4,432	233	286	540	5,098	769	28
Michigan	18,680	6,770	771	340	815	8,952	989	43
Minnesota	19,938	3,985	502	261	796	12,952	1,391	51
Mississippi	6,620	3,113	190	114	400	2,473	322	8
Missouri	20,367	5,349	877	364	1,452	11,109	1,165	51
Montana	3,313	1,272	88	125	178	1,311	332	7
Nebraska	5,294	1,531	319	164	434	2,185	645	16
Nevada	11,915	3,428	1,236	158	294	6,385	405	9
New Hampshire	6,984	4,425	711	109	183	1,245	299	12
New Jersey	25,906	3,666	578	136	978	20,124	404	20
New Mexico	17,370	2,964	652	196	386	12,948	216	8
New York	24,794	4,795	252	518	1,768	16,348	1,071	42
North Carolina	29,030	6,561	1,994	305	966	18,550	634	20
North Dakota	2,927	1,019	1,014	60	125	546	159	4
Ohio	53,310	11,954	3,250	656	2,385	33,487	1,534	44
Oklahoma	16,279	6,734	1,742	341	910	5,608	925	19
Oregon	18,705	4,834	3,646	642	888	7,331	1,336	28
Pennsylvania	52,358	13,519	3,185	700	1,439	32,000	1,410	105
Rhode Island	1,930	416	15	34	96	1,321	42	6
South Carolina	14,401	2,901	560	197	1,165	9,011	553	14
South Dakota	2,942	1,060	115	72	141	1,216	330	8
Tennessee	26,689	6,472	2,402	284	2,393	13,912	1,210	16
Texas	84,019	19,415	16,555	1,344	4,055	39,359	3,197	94
Utah	12,347	6,514	678	110	350	4,466	229	0
Vermont	3,086	1,029	44	44	56	1,706	204	3
Virginia	62,513	14,049	2,864	441	2,683	40,821	1,640	15
Washington	20,817	2,362	706	478	582	15,363	1,301	25
West Virginia	4,904	1,791	398	183	219	1,946	364	3
Wisconsin	19,347	4,130	1,109	213	904	12,341	642	8
Wyoming	76,800	1,298	140	83	337	74,651	278	13
Totals	1,244,058	278,958	85,996	15,379	57,543	764,645	40,457	1,080

Source: Bureau of Alcohol, Tobacco and Firearms.

[a] Destructive devices generally consist of certain explosives or incendiary devices, but they also include certain specific firearms, such as the USAS-12 and the Streetsweeper.

[b] The term "any other weapon" includes any weapon or device capable of being concealed on the person that can be discharged through the energy of an explosive; a pistol or revolver having a barrel with a smooth bore that can fire a fixed shotgun shell; weapons with combination shotgun and rifle barrels of a certain size from which only a single discharge can be made from either barrel without manual reloading; or any such weapon that can be readily restored to so fire.

[c] "Other" includes firearms that meet the legal definition of firearms under the National Firearms Act but cannot be categorized as machineguns, silencers, short-barreled rifles or shotguns, destructive devices, or any other weapon.

Exhibit 10. National Firearms Act Special Occupational Taxpayers (FY 1980 - FY 2001)

Fiscal Year	Special Occupational Taxpayers [a]	Percent Change from FY 1980
1980	920	
1981	1,192	30%
1982	1,758	91%
1983	2,306	151%
1984	2,678	191%
1985	2,696	193%
1986	3,297	258%
1987	5,427	490%
1988	3,673	299%
1989	2,977	224%
1990	2,827	207%
1991	2,775	202%
1992	2,754	199%
1993	2,733	197%
1994	2,684	192%
1995	2,468	168%
1996	2,283	148%
1997	2,499	172%
1998	2,283	148%
1999	2,521	174%
2000	2,668	190%
2001	2,490	171%

Source: Bureau of Alcohol, Tobacco and Firearms.

[a] Special occupational taxpayers are persons who wish to manufacture, import, or deal in firearms as defined in the NFA. Special occupational taxpayers must (1) be properly licensed as a Federal firearms licensee; (2) have an employer identification number (even if the licensee has no employees); and (3) pay the special occupational tax required of those manufacturing, importing, or dealing in NFA weapons.

Exhibit 11. National Firearms Act Special Occupational Taxpayers (as of January 2001) by State

State	Total	Importers	Manufacturers	Dealers
Alabama	62	3	23	36
Alaska	17	0	2	15
Arizona	151	4	47	100
Arkansas	33	1	13	19
California	112	6	32	74
Colorado	36	1	7	28
Connecticut	69	5	21	43
Delaware	0	0	0	0
District of Columbia	1	0	0	1
Florida	217	12	69	136
Georgia	94	6	21	67
Hawaii	1	0	0	1
Idaho	34	0	22	12
Illinois	77	5	17	55
Indiana	63	0	17	46
Iowa	11	1	3	7
Kansas	25	2	6	17
Kentucky	37	1	9	27
Louisiana	52	1	10	41
Maine	16	2	8	6
Maryland	60	3	22	35
Massachusetts	44	0	19	25
Michigan	68	8	12	48
Minnesota	55	5	34	16
Mississippi	26	0	5	21
Missouri	87	3	36	48
Montana	12	0	1	11
Nebraska	12	0	6	6
Nevada	61	4	24	33
New Hampshire	35	1	10	24
New Jersey	14	0	2	12
New Mexico	30	1	9	20
New York	16	0	7	9
North Carolina	90	1	19	70
North Dakota	5	0	2	3
Ohio	127	2	39	86
Oklahoma	45	0	14	31
Oregon	75	0	24	51
Pennsylvania	132	2	37	93
Rhode Island	4	0	0	4
South Carolina	17	1	15	1
South Dakota	13	0	1	12
Tennessee	72	4	25	43
Texas	248	4	46	198
Utah	22	3	11	8
Vermont	11	2	5	4
Virginia	92	11	27	54
Washington	22	2	14	6
West Virginia	21	0	10	11
Wisconsin	34	0	8	26
Wyoming	10	1	3	6
Total	2,668	108	814	1,746

Source: Bureau of Alcohol, Tobacco and Firearms.

Exhibit 12. Federal Firearms Licensees Total (FY 1975 - FY 2001)

Fiscal Year	Total	Change from prior year	Dealer	Pawn-broker	Collector	Manufacturer Ammuni-tion	Firearms	Importer	Destructive Device Dealer	Armor Piercing Ammunition Manufacturer	Importer
1975	161,927		146,429	2,813	5,211	6,668	364	403	9	23	7
1976	165,697	2.3%	150,767	2,882	4,036	7,181	397	403	4	19	8
1977	173,484	4.7%	157,463	2,943	4,446	7,761	408	419	6	28	10
1978	169,052	-2.6%	152,681	3,113	4,629	7,735	422	417	6	35	14
1979	171,216	1.3%	153,861	3,388	4,975	8,055	459	426	7	33	12
1980	174,619	2.0%	155,690	3,608	5,481	8,856	496	430	7	40	11
1981	190,296	9.0%	168,301	4,308	6,490	10,067	540	519	7	44	20
1982	211,918	11.4%	184,840	5,002	8,602	12,033	675	676	12	54	24
1983	230,613	8.8%	200,342	5,388	9,859	13,318	788	795	16	71	36
1984	222,443	-3.5%	195,847	5,140	8,643	11,270	710	704	15	74	40
1985	248,794	11.8%	219,366	6,207	9,599	11,818	778	881	15	85	45
1986	267,166	7.4%	235,393	6,998	10,639	12,095	843	1,035	16	95	52
1987	262,022	-1.9%	230,888	7,316	11,094	10,613	852	1,084	16	101	58
1988	272,953	4.2%	239,637	8,261	12,638	10,169	926	1,123	18	112	69
1989	264,063	-3.3%	231,442	8,626	13,536	8,345	922	989	21	110	72
1990	269,079	1.9%	235,684	9,029	14,287	7,945	978	946	20	117	73
1991	276,116	2.6%	241,706	9,625	15,143	7,470	1,059	901	17	120	75
1992	284,117	2.9%	248,155	10,452	15,820	7,412	1,165	894	15	127	77
1993	283,925	-0.1%	246,984	10,958	16,635	6,947	1,256	924	15	128	78
1994	250,833	-11.7%	213,734	10,872	17,690	6,068	1,302	963	12	122	70
1995	191,495	-23.7%	158,240	10,155	16,354	4,459	1,242	842	14	118	71
1996	135,794	-29.1%	105,398	9,974	14,966	3,144	1,327	786	12	117	70
1997	107,554	-20.8%	79,285	9,956	13,512	2,451	1,414	733	13	118	72
1998	105,536	-1.9%	75,619	10,176	14,875	2,374	1,546	741	12	125	68
1999	103,942	-1.5%	71,290	10,035	17,763	2,247	1,639	755	11	127	75
2000	103,157	-0.8%	67,479	9,737	21,100	2,112	1,773	748	12	125	71
2001	102,913	-0.2%	63,845	9,199	25,145	1,950	1,841	730	14	117	72

Source: Bureau of Alcohol, Tobacco and Firearms, National Licensing Center. Data is based on active firearms licenses and related statistics as of the end of the fiscal year.

Exhibit 13. Federal Firearms Licensees by State, Number, and Rate per 100,000 Population (2001)

State	Population	Number of FFLs as of 2001	FFLs per 100,000 Population
Alabama	4,332,379	1,911	44
Alaska	607,582	1,237	204
Arizona	5,020,782	1,900	38
Arkansas	2,599,491	1,504	58
California	33,051,895	6,093	18
Colorado	4,198,307	1,879	45
Connecticut	3,297,626	1,222	37
Delaware	759,017	197	26
District of Columbia	536,497	12	2
Florida	15,593,435	4,438	28
Georgia	7,952,628	2,962	37
Hawaii	1,175,755	203	17
Idaho	1,262,458	1,148	91
Illinois	12,097,507	3,547	29
Indiana	5,902,333	2,544	43
Iowa	2,822,157	1,685	60
Kansas	2,606,470	1,486	57
Kentucky	3,926,964	1,931	49
Louisiana	4,333,010	1,827	42
Maine	1,240,011	767	62
Maryland	5,162,430	1,482	29
Massachusetts	6,127,881	2,002	33
Michigan	9,688,556	4,323	45
Minnesota	4,783,596	2,590	54
Mississippi	2,749,243	1,540	56
Missouri	5,433,154	3,781	70
Montana	877,432	2,997	342
Nebraska	1,660,444	1,020	61
Nevada	1,964,582	847	43
New Hampshire	1,200,247	740	62
New Jersey	8,219,529	631	8
New Mexico	1,782,739	983	55
New York	18,395,994	3,687	20
North Carolina	7,795,432	2,997	38
North Dakota	618,569	653	106
Ohio	11,054,019	3,987	36
Oklahoma	3,338,278	1,936	58
Oregon	3,343,908	2,232	67
Pennsylvania	11,847,752	4,867	41
Rhode Island	1,009,503	296	29
South Carolina	3,876,975	1,226	32
South Dakota	726,426	647	89
Tennessee	5,541,336	2,401	43
Texas	20,290,713	7,535	37
Utah	2,192,690	862	39
Vermont	588,067	575	98
Virginia	6,847,117	2,878	42
Washington	5,757,739	2,013	35
West Virginia	1,765,197	1,379	78
Wisconsin	5,207,717	2,466	47
Wyoming	479,699	774	161
Total	273,643,268	104,840	38

Source: Population data, Census Bureau; FFL data, Bureau of Alcohol, Tobacco and Firearms

Exhibit 13. Federal Firearms Licensees by State, Number, and Rate per 100,000 Population (2001)

Population	Number of FFLs	FFLs per 100,000

E-13

Exhibit 14. License Applications and Application Inspections (FY 1990 - FY 2001)

Fiscal Year	New Applicants	Renewals	Application Inspections
1990	34,336	61,536	3,358
1991	34,567	57,327	4,000
1992	37,085	58,873	3,582
1993	41,545	66,811	4,701
1994	25,393	37,079	2,462
1995	7,777	19,541	4,815
1996	8,461	34,304	6,385
1997	6,188	30,660	6,430
1998	6,881	26,042	8,959
1999	8,581	31,978	5,351
2000	10,698	22,333	4,016
2001	11,161	24,710	5,497

Source: Bureau of Alcohol, Tobacco and Firearms.

Exhibit 15. Actions on Federal Firearms License Applications (FY 1975 - FY 2001)

Fiscal Year	Original Application				Renewal Applications			
	Processed	Denied	Withdrawn [a]	Abandoned [b]	Processed	Denied	Withdrawn	Abandoned
1975	29,183	150	1,651	...	138,719	273	334	...
1976	29,511	209	2,077	...	138,050	261	436	...
1977	32,560	216	1,645	...	136,629	207	409	...
1978	29,531	151	1,015	414	139,383	168	141	449
1979	32,678	124	432	433	143,021	93	240	942
1980	36,052	96	601	661	143,527	31	336	800
1981	41,798	85	742	329	152,153	16	385	495
1982	44,745	52	580	370	161,390	12	332	350
1983	49,669	151	916	649	163,386	48	514	700
1984	39,321	98	706	833	163,950	23	449	825
1985	37,385	103	666	598	52,768	9	226	307
1986	42,842	299	698	452	47,648	14	135	181
1987	36,835	121	874	458	61,596	38	428	225
1988	32,724	30	506	315	52,738	19	422	182
1989	34,318	34	561	360	54,892	14	1,456	215
1990	34,336	46	893	404	61,536	29	48	63
1991	34,567	37	1,059	685	57,327	15	82	106
1992	37,085	57	1,337	611	58,873	4	26	88
1993	41,545	343	6,030	1,844	66,811	53	1,187	683
1994	25,393	136	4,480	3,917	37,079	191	1,128	969
1995	7,777	49	1,046	1,180	19,541	65	1,077	1,254
1996	8,461	58	1,061	629	34,304	99	2,700	980
1997	7,039	24	692	366	30,660	144	2,185	801
1998	7,090	19	621	352	26,042	65	689	509
1999	8,581	23	48	298	31,978	63	698	539
2000	10,698	6	447	91	22,333	27	118	135
2001	11,161	3	403	114	24,710	28	489	153

Source: Bureau of Alcohol, Tobacco and Firearms.

[a] An application can be withdrawn by an applicant at any time prior to the issuance of a license.

[b] If ATF cannot locate an applicant during an attempted application inspection or cannot obtain needed verification data, then the application will be abandoned.

Exhibit 16. Federal Firearms Licensees and Compliance Inspections (FY 1969 - FY 2001)

Fiscal Year	Licensees	Inspections	Percent Inspected
1969 [a]	86,598	47,454	54.7
1970	138,928	21,295	15.3
1971	149,212	32,684	21.9
1972	150,215	31,164	20.7
1973	152,232	16,003	10.5
1974	158,753	15,751	10.0
1975	161,927	10,944	6.7
1976	165,697	15,171	9.1
1977	173,484	19,741	11.3
1978	169,052	22,130	13.1
1979	171,216	14,744	8.6
1980	174,619	11,515	6.5
1981	190,296	11,035	5.7
1982	211,918	1,829	0.8
1983	230,613	2,662	1.1
1984	222,443	8,861	3.9
1985	248,794	9,527	3.8
1986	267,166	8,605	3.2
1987	262,022	8,049	3.1
1988	272,953	9,283	3.4
1989	264,063	7,142	2.7
1990	269,079	8,471	3.1
1991	276,116	8,258	3.0
1992	284,117	16,328	5.7
1993	283,925	22,330	7.9
1994	250,833	20,067	8.0
1995	187,931	13,141	7.0
1996	135,794	10,051	7.4
1997	107,554	5,925	5.5
1998	105,536	5,043	4.8
1999	103,942	9,004	8.7
2000 [b]	103,658	3,640	3.5
2001	102,913	3,677	3.6

Source: Bureau of Alcohol, Tobacco and Firearms.

[a] Statistics from 1969 to 1971 reflect both application and compliance inspections conducted.

[b] Overall numbers of inspections declined as the result of the Bureau's comprehensive focused inspection project.

www.ingramcontent.com/pod-product-compliance
Lightning Source LLC
Chambersburg PA
CBHW081415170526
45166CB00010B/3354